U.S. SEISMIC ZONES

THE NEW MADRID SEISMIC ZONE

GREG ROZA

PowerKiDS press

Published in 2026 by The Rosen Publishing Group, Inc.
2544 Clinton Street, Buffalo, NY 14224

First Edition

Editor: Greg Roza
Book Design: Michael Flynn

Photo Credits: Cover anthony heflin/Shutterstock.com; p. 5 zelvan/Shutterstock.com; p. 7 Rainer Lesniewski/Shutterstock.com; p. 9 stihii/Shutterstock.com; p. 10 https://commons.wikimedia.org/wiki/File:NMSZBig.gif; p. 11 https://commons.wikimedia.org/wiki/File:New_Madrid_Seismic_Zone_activity_1974-2011.svg; p. 12 Robert Biedermann/Shutterstock.com; p. 13 https://commons.wikimedia.org/wiki/File:Pemiscot-ditch-mo.jpg; p. 15 https://commons.wikimedia.org/wiki/File:Mississippi_Embayment_relief_map_2.svg; p. 17 https://commons.wikimedia.org/wiki/File:Soil_liquefaction_(24399502593).jpg; p. 19 Everett Collection/Shutterstock.com; p. 21 https://commons.wikimedia.org/wiki/File:The_Great_Earthquake_at_New_Madrid_(15_September_1811)_-_DPLA_-_c35a97038658891336099 29fa82173dd.jpg; p. 23 The Reading Room / Alamy Stock Photo; p. 25 (main) kwan tse/Shutterstock.com; p. 25 (inset) https://commons.wikimedia.org/wiki/File:PSM_V69_D087_Cypress_growth_in_reelfoot_lake_tenn_after_new_madrid_earthquake.png; p. 27 UPI / Alamy Stock Photo; p. 29 RP Images / Alamy Stock Photo.

Cataloging-in-Publication Data

Names: Roza, Greg.
Title: The New Madrid Seismic Zone / Greg Roza.
Description: Buffalo, New York : PowerKids Press, 2026. | Series: U.S. seismic zones | Includes glossary and index.
Identifiers: ISBN 9781499453355 (pbk.) | ISBN 9781499453362 (library bound) | ISBN 9781499453379 (ebook)
Subjects: LCSH: Faults (Geology)--Juvenile literature. | Plate tectonics--Juvenile literature. | Earthquakes--Missouri--Juvenile literature. | Seismology--Juvenile literature.
Classification: LCC QE606.R69 2026 | DDC 551.8'72--dc23

Manufactured in the United States of America

CONTENTS

SEISMIC ODDITY

Located in the southeast United States, the New Madrid seismic zone (NMSZ) is somewhat of an oddity in the world of geology. It's a region that has experienced powerful earthquakes in the not-too-distant past, and many scientists believe it will experience more in the not-too-distant future. However, its location is not exactly typical when it comes to seismic zones.

The NMSZ covers parts of northeastern Arkansas, southwestern Kentucky, southeastern Missouri, and northwestern Tennessee. This area is in the flood plain of the Mississippi River. Major cities that might be affected by earthquakes that occur in the NMSZ include Little Rock, Arkansas; St. Louis, Missouri; and Memphis, Tennessee. The zone follows the position of the Mississippi River, which we will soon find has a lot to do with the seismic activity in the area.

Most seismic zones feature mountains, volcanoes, visible fault lines, and ancient lava plains. The NMSZ, however, has none of those features. Why is this area so different from other, more typical seismic zones? Let's find out!

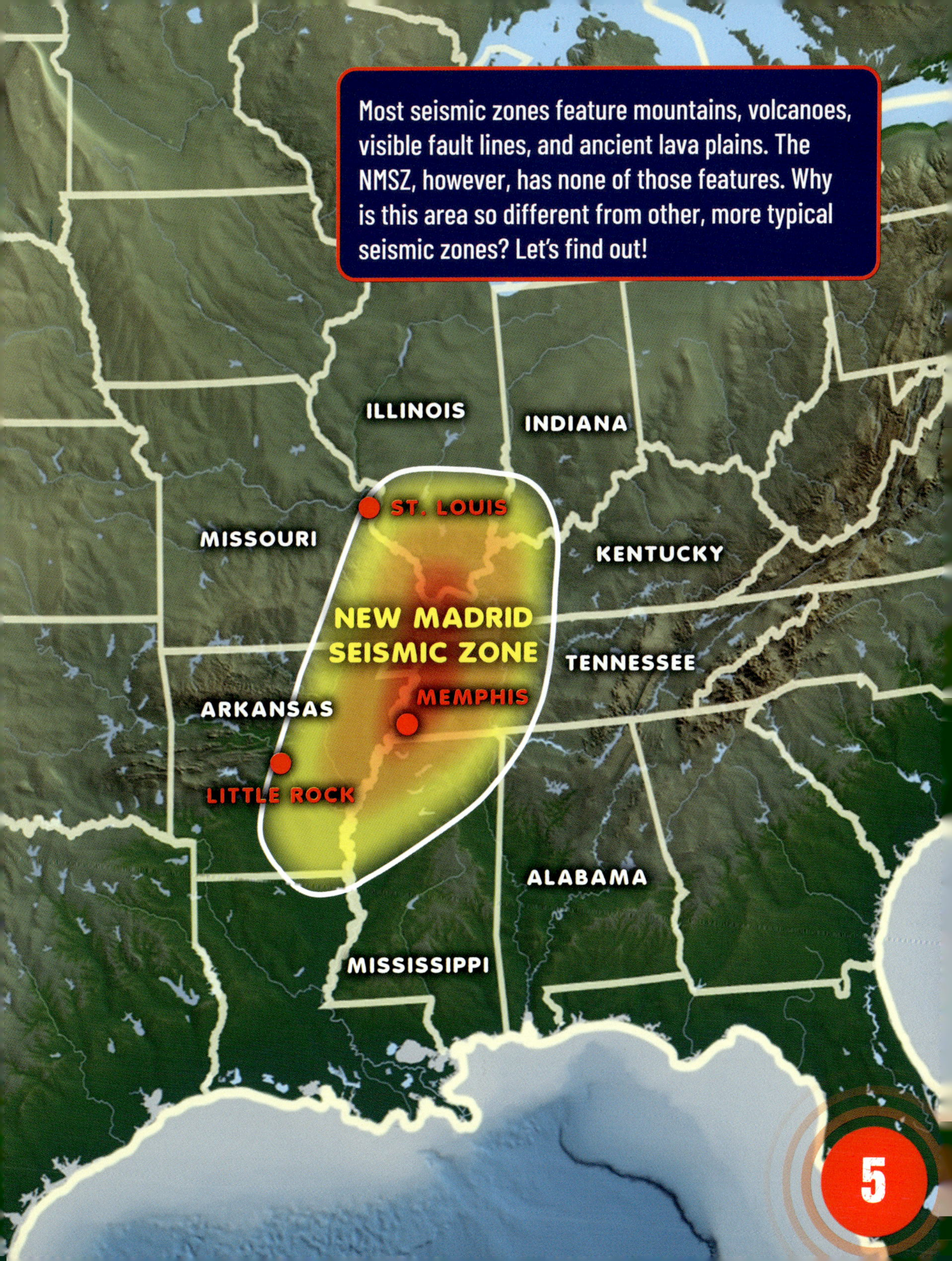

WHAT MAKES THE EARTH MOVE?

Seismic activity, or earthquakes, is caused by the movement of enormous sections of Earth's crust called plates. These plates drift very slowly. They bump and crash into each other, often building up pressure between the plates. In time—which can be hundreds or thousands of years—the plates suddenly move, releasing the pressure that has built up as the they settle into new positions. This result is an earthquake.

Most seismic areas are along fault lines, and most faults are the spaces where plates meet. The most active seismic area on Earth is called the Ring of Fire, which is a 25,000-mile (40,234 km) path of fault lines surrounding the Pacific Ocean. It's called a ring of "fire" because of the number of volcanoes, both active and **dormant**, along the path.

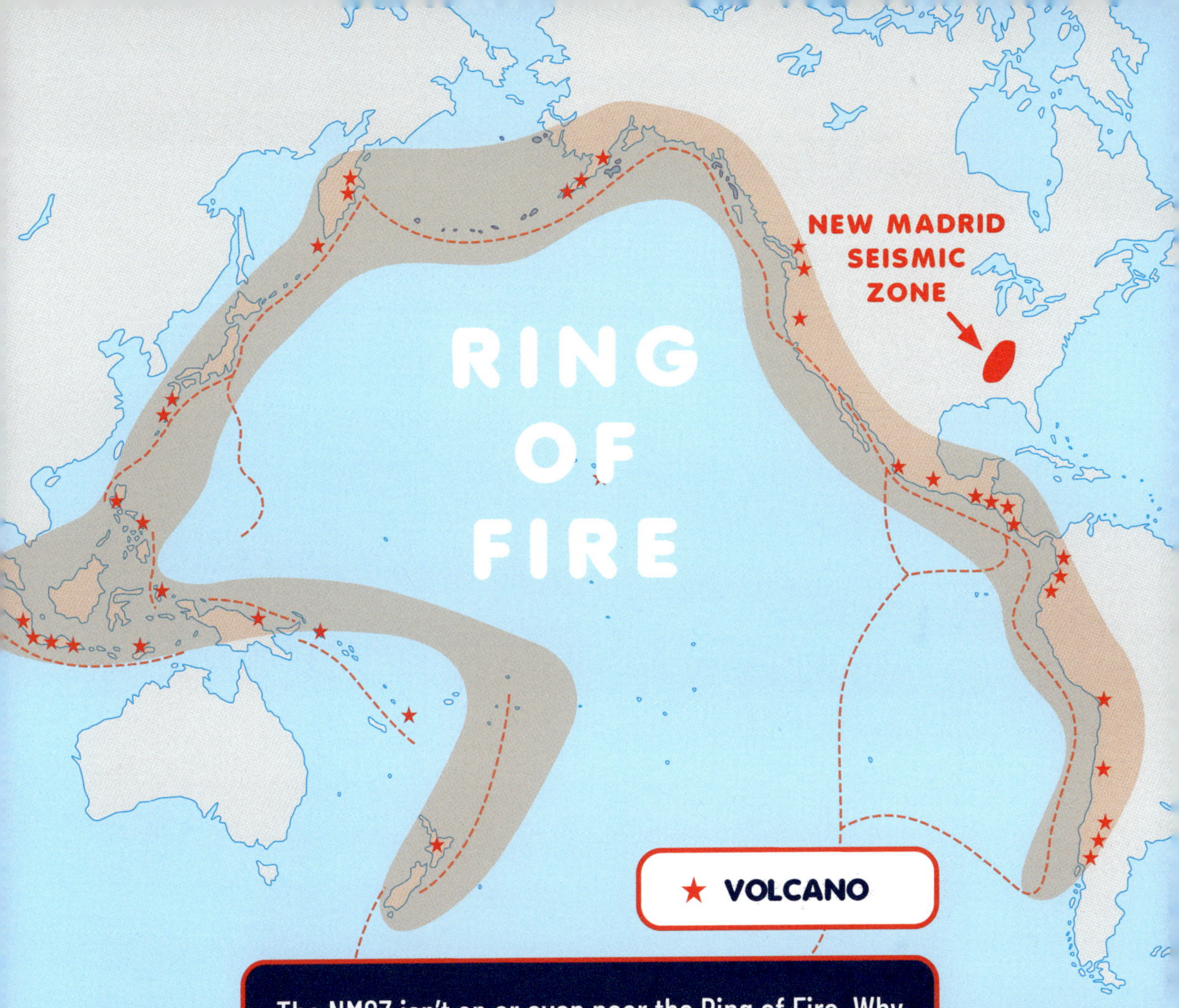

The NMSZ isn't on or even near the Ring of Fire. Why is a notable seismic zone, with faults of its own, so far away from the faults between Earth's plates?

EARTH-SHAKING FACT!

About 200 million years ago, most of Earth's land was grouped together in a **supercontinent** scientists call Pangea. Over time, Pangea broke into smaller sections. These smaller landmasses drifted into the positions they are now in—Earth's seven continents.

THE REELFOOT RIFT

Geologists believe that supercontinents have broken up and formed new supercontinents many times. When a supercontinent starts to break up, faults form across the landmass. The land on either side of a fault may drift apart, but sometimes Earth's forces stop them.

A supercontinent named Rodinia formed more than 1 billion years ago. "Shortly" after, it began to break up about 750 million years ago. About 600 million years ago, a fault began to form on a part of Rodinia but it failed to create a full break. This resulted in a weakened area of land in the middle of the plate. In time, another landmass, the North American plate, covered the weakened part of Rodinia, **embedding** it under newer layers of earth. Today, geologists call this weakened area the Reelfoot Rift.

A rift is a weakened area on Earth's crust that forms where two landmasses are pulling away from each other. Rifts form cracks in the crust called faults. The spaces between Earth's plates are also called faults.

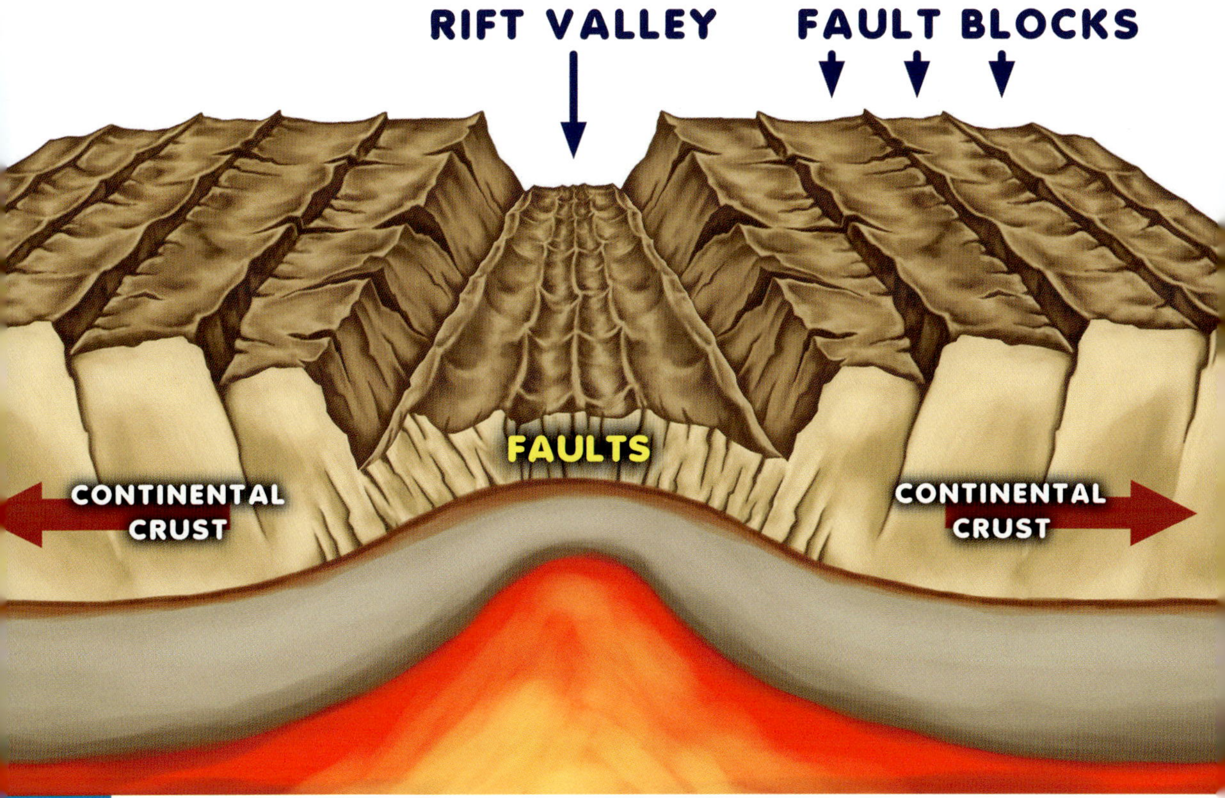

BUT . . . WHY?

Most scientists think the Reelfoot Rift is the main cause of the NMSZ, but it's more complicated than that. Scientists have discovered similar weakened areas beneath the North American plate, including a **significant** one beneath Lake Superior. However, those areas lack seismic activity. Geologists continue to study the NMSZ to hunt out factors that could explain why the area experiences earthquakes.

Many seismic zones are easy to study because their results are aboveground and plain to see. The frequent lava flows that occur in Iceland are a perfect example! The Reelfoot Rift, however, is buried beneath 100 to 200 feet (30.5 to 61 m) of **sediment** left behind by the Mississippi River. This makes it very hard to study!

Geologists are constantly monitoring the NMSZ and mapping its seismic activity. Minor quakes (less than a 2.0 **magnitude**) happen on average every other day. People usually don't notice an earthquake of this magnitude, but modern technology catches every single **tremor**. The information geologists have amassed over the years shows that Reelfoot Rift has four zig-zagging faults that stretch for about 125 miles (201 km) between northeast Arkansas, southeast Missouri, and northwest Tennessee.

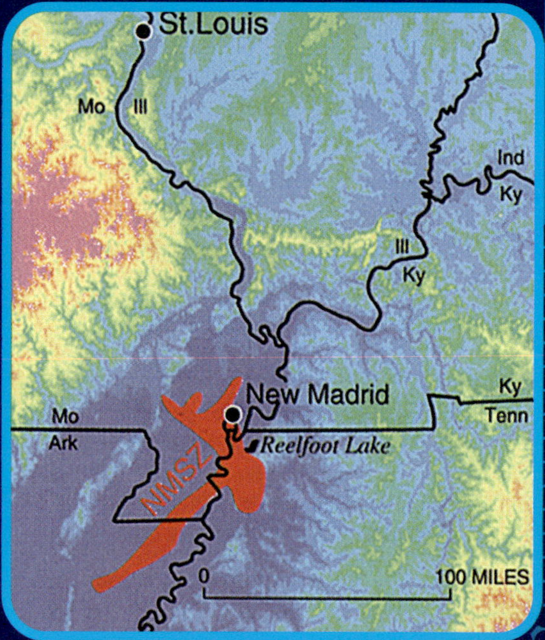

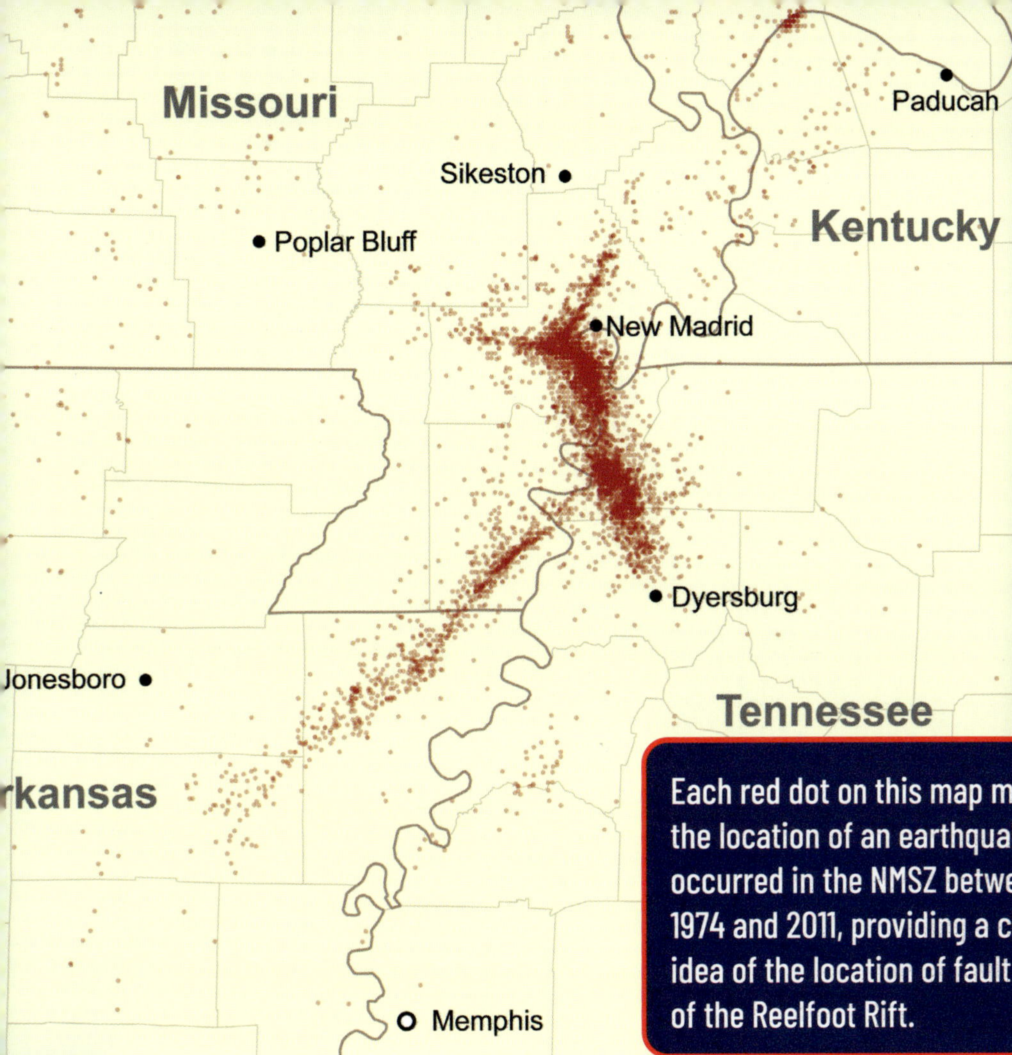

Each red dot on this map marks the location of an earthquake that occurred in the NMSZ between 1974 and 2011, providing a clear idea of the location of fault lines of the Reelfoot Rift.

THE WABASH VALLEY SEISMIC ZONE

The Wabash Valley seismic zone is an even smaller area to the northeast of the NMSZ. It is around the Wabash River, which forms part of the border between the states of Illinois and Indiana. Scientists believe it was formed by the same forces that created the NMSZ. This area has experienced earthquakes for the past 20,000 years. On April 18, 2008, one of the largest earthquakes to hit the state of Illinois in recorded history occurred near Mount Carmel. It resulted in minor building damage.

THEORIES AND EVIDENCE

Major seismic zones commonly include volcanic activity. The Hawaiian Islands are the volcanic peaks of massive seamounts created by a deep-sea hotspot on the ocean floor. The Cascade Range is the result of two of Earth's plates pushing together, which creates both seismic and volcanic **phenomena**. However, the NMSZ is a flat valley that does not experience volcanic eruptions.

The Reelfoot Rift is largely made up of igneous rock. This means it was once lava that came out of a volcano. The newer layer of crust over the rift is largely sedimentary rock. Sedimentary rock is formed by the deposit and **accumulation** of eroded sediments. Rather than pushing land up to form peaks, seismic activity in the NMSZ results in ground weakness, landslides, and sinking land.

The area of Missouri where the NMSZ is located is called the Bootheel because of its odd shape. It has flat farmland and drainage canals, such as this one near Bragg City.

JOHN HARDEMAN WALKER

In 1810, when John Hardeman Walker was just 16, he moved from Tennessee to the area then called the Missouri Territory. After the earthquakes of 1811–1812, many people left. Walker stayed and became rich and powerful raising cattle. When the U.S. government established the states of Missouri and Arkansas, Walker asked Congress to keep the area where he lived as a part of Missouri instead of Arkansas. Congress listened to Walker, and in 1821, Missouri—with Walker's Bootheel—became the 24th state.

The NMSZ sits at the north end of a geological area known as the Mississippi embayment. This is a triangle-shaped area of land made up of the Mississippi **Alluvial** Plain and surrounding area. The embayment covers all of Louisiana and reaches the Gulf of Mexico. The land of the embayment is flatter and lower than surrounding areas. After hundreds of years, the Mississippi River spread sediment over the flood plain and much of the land is or was swampland and forests. Manmade canals and drainage ditches helped dry out the land for farming.

The Mississippi River has cut down into Earth's crust for 40 million years. Some scientists think this downcutting action has resulted in weaknesses in the Reelfoot Rift, leading to seismic activity.

EARTH-SHAKING FACT!

Some scientists believe the NMSZ might be over a hotspot below Earth's crust, similar to the hotspot that forms the Hawaiian islands. A hotspot is an area of the mantle that is hotter than surrounding areas. A hotspot can cause weaknesses in the crust above it, as well as seismic and volcanic activity.

Before European settlement, the Mississippi Alluvial Plain was the largest wetland and forest ecosystem in North America.

SAND BOILS

The NMSZ doesn't have volcanic peaks, but it does have volcanoes . . . sand volcanoes! Also called sand boils, they aren't volcanoes, but they look and act somewhat like volcanoes. They also help scientists discover what happened in the NMSZ over the course of thousands of years.

As we've learned, the NMSZ is largely made up of sedimentary rock and alluvial sediment. Earthquakes have an interesting effect on sand called liquefaction. Liquefaction occurs when loosely packed, water-logged soil or sand becomes unstable during seismic activity. The soil or sand can seem to flow like water! This can cause buildings to topple.

EARTH-SHAKING FACT!

Sand boils that formed during the 1811–1812 earthquakes covered a very large area—about 6,460 square miles (16,731 sq km). Many are still visible, especially closer to the epicenters near New Madrid. Scientists have discovered that some of the sand boils are from earlier, prehistoric earthquakes.

Scientists have uncovered hardened sand blows that prove sizeable earthquakes struck the Mississippi River valley around 2530 BCE, 300 BCE, 900 CE, and 1450 CE.

When seismic activity hits an area of loose, wet sand, it can cause the sand and water to form a **geyser**. This creates sand boils, which are wide cones of sand on Earth's surface.

MAJOR QUAKES

The NMSZ was once a vast stretch of forest and wetlands. Even today, much of it features woodland and swamps. It is crisscrossed by numerous creeks and rivers and includes a **diverse** range of wildlife. The Mississippi River has shaped the land and wildlife in the area for millions of years.

In the late 1700s, European settlers started clearing land for farms and ranches. However, the peaceful farmland became the location of some of the largest earthquakes ever to hit North America! Scientists estimate that three earthquakes during this time were each between a magnitude of 7 and 8. An **aftershock** of the December 1811 quake was also in this range. This is powerful enough to tear down even more modern buildings than those they had in New Madrid and surrounding area at the time.

This painting appeared in an 1867 book about life in Missouri. It shows a happy family living a simple life. Do you think they would be prepared for an earthquake?

EARTH-SHAKING FACT!

Researchers still aren't sure how many people died during the 1811–1812 earthquakes. Even though the quakes were very powerful, the death toll was likely low due to the small, rural population in the area.

December 16, 1811, 2:15 a.m.

The first earthquake struck in the middle of the night, but the epicenter was in an unpopulated area in northeast Arkansas. It caused some damage to buildings. Tremors were felt in Chickasaw Bluffs, which is where modern-day Memphis, Tennessee, is located. North of the epicenter, the town of Little Prairie, Missouri, was heavily damaged by liquefaction.

January 23, 1812, 9:15 a.m.

The epicenter for the second earthquake was near New Madrid and resulted in more damage to buildings. There were numerous reports of ground warping, landslides, and soil liquefaction.

February 7, 1812, 3:45 a.m.

The epicenter for the third earthquake was once again near New Madrid, and this time it destroyed the town. There were reports of building damage in St. Louis, Missouri, 165 miles (266 km) away. There were widespread ground weaknesses and landslides. New faults created temporary waterfalls on the Mississippi River. This quake also created Reelfoot Lake.

EARTH-SHAKING FACT!

On December 16, 1811, an aftershock struck about five hours after the main shock. It had a similar magnitude to the main shock. The NMSZ experienced hundreds of aftershocks that lasted into 1813, although none were as powerful as the first one.

This drawing of destruction being caused by the NMSZ earthquake was first published in 1851.

FIRST-PERSON ACCOUNTS

Much of what scientists know about the three earthquakes comes from people who witnessed the destruction caused by them. Eliza Bryan told her story in a letter she sent to her friend, Reverend Lorenzo Dow. Dow then shared the report with others.

In her letter, Bryan describes the tremors she felt during the December 16 aftershock: "The hardest shock of all [was] when the waters of the river gathered up like a mountain, rising 15 to 20 feet perpendicularly (upwards) and then receding (going back) within its banks with such violence that it took with it whole groves of young cottonwoods which edged its borders. Fissures (cracks) in the earth vomited forth sand and water, some closing again immediately."

Firmin La Roche was the leader of a group of flatboats. In his report, he said: "... a wave [so big] came up the river that I never seen one like it at sea. It carried us back north, upstream, for more than a mile, and the water spread out upon the banks, covering maybe 3 or 4 miles inland."

JOHN BRADBURY

English scientist John Bradbury traveled the U.S. Midwest in the early 1800s. His first-person account of the 1811–1812 earthquakes was the only one from a scientist. Bradbury and other men survived the earthquake while on a ship on the Mississippi River. He described the scene on land: "I went ashore and found the chasm (deep crack) really frightful, as it was not less than 4 feet in width, and besides the bank had sunk at least 2 feet. I . . . examined to determine (the chasm's) length and concluded that it could not be less than 80 yards . . . ."

AFTERMATH

The earthquakes of 1811–1812 permanently scarred the land with cracks, sunken land, uplifted land, and sand boils. Lakes and swamps were both created and destroyed. Landslides made creeks and rivers wider and sometimes changed their courses. Several islands in the Mississippi River disappeared. The New Madrid cemetery had been washed away!

In Arkansas, some areas sank so much that it resulted in flooding every spring. This made farming in the area almost impossible until 1893 when the first **levies** were built to hold back floodwaters. The water level of St. Francis Lake is believed to have risen up to 29.5 feet (9 m)!

In Tennessee, the town of Tiptonville reached a new height. At the same time, ground weakness east of Tiptonville resulted in the creation of Reelfoot Lake.

EARTH-SHAKING FACT!

The earthquakes caused land in Tennessee east of the Mississippi River to sink. As Firmin La Roche's report shows, the current of the river actually went backward for a time. The backwards flowing water splashed out of the river and into the newly sunken area, creating Reelfoot Lake. Today, the area is a popular place to fish and birdwatch.

The first earthquake was felt from Canada to New Orleans. In the end, between 3,000 to 5,000 square miles (7,800 to 13,000 square km) of land was left visibly scarred.

REELFOOT LAKE TODAY

REELFOOT LAKE 1800S

SINCE THEN

Thousands of aftershocks followed the 1811–1812 quakes over a period of several years. Some were strong enough to be felt, and others were much weaker. Other notable quakes were recorded in 1843 and 1895. The 1895 quake had a magnitude that was close to 7 and caused widespread building damage. The epicenter for the 1895 quake was beneath Charleston, Missouri, and damaged nearly all of the town's buildings. Building damage was reported in Memphis, St. Louis, and Evansville, Indiana.

The largest NMSZ earthquake in the 20th century occurred in 1968 near the town of Dale, Illinois. This area is actually considered part of the Wabash River seismic zone. This earthquake was felt in 23 states and in Ontario, Canada. It caused some building damage in nearby communities.

The continuing seismic activity in the NMSZ, regardless of how small, is a sure sign that the area will likely experience more major quakes in the future.

IBEN BROWNING

Browning was a business consultant and self-proclaimed weather scientist, even though he had no training (he had a PhD in zoology, or the study of animals). He predicted that a powerful earthquake would strike the New Madrid area on December 3, 1990. His prediction caused a panic! The Red Cross gave out 230,000 earthquake preparedness kits. Some communities closed their schools on December 3. However, no quake occurred. Actual scientists were not surprised.

THE NEXT BIG ONE?

Seismic activity in the NMSZ rises and falls. It increased in late 2024, including 10 quakes (less than a magnitude of 3) in December alone. Many scientists believe there is a strong chance of another earthquake of similar magnitude to the 1811–1812 earthquakes occurring in the next 50 years. Such a quake would be sure to cripple cities such as Memphis and Little Rock. It could also disturb trade routes, including the Mississippi River and railroads. However, not all scientists are sure that a large quake will happen in the near future.

Although Iben Browning's prediction was wrong, it did spark communities to become better prepared for potential seismic activity. The Central United States Earthquake Consortium (CUSEC), established in 1984, has worked hard to prepare Midwesterners for the next big quake.

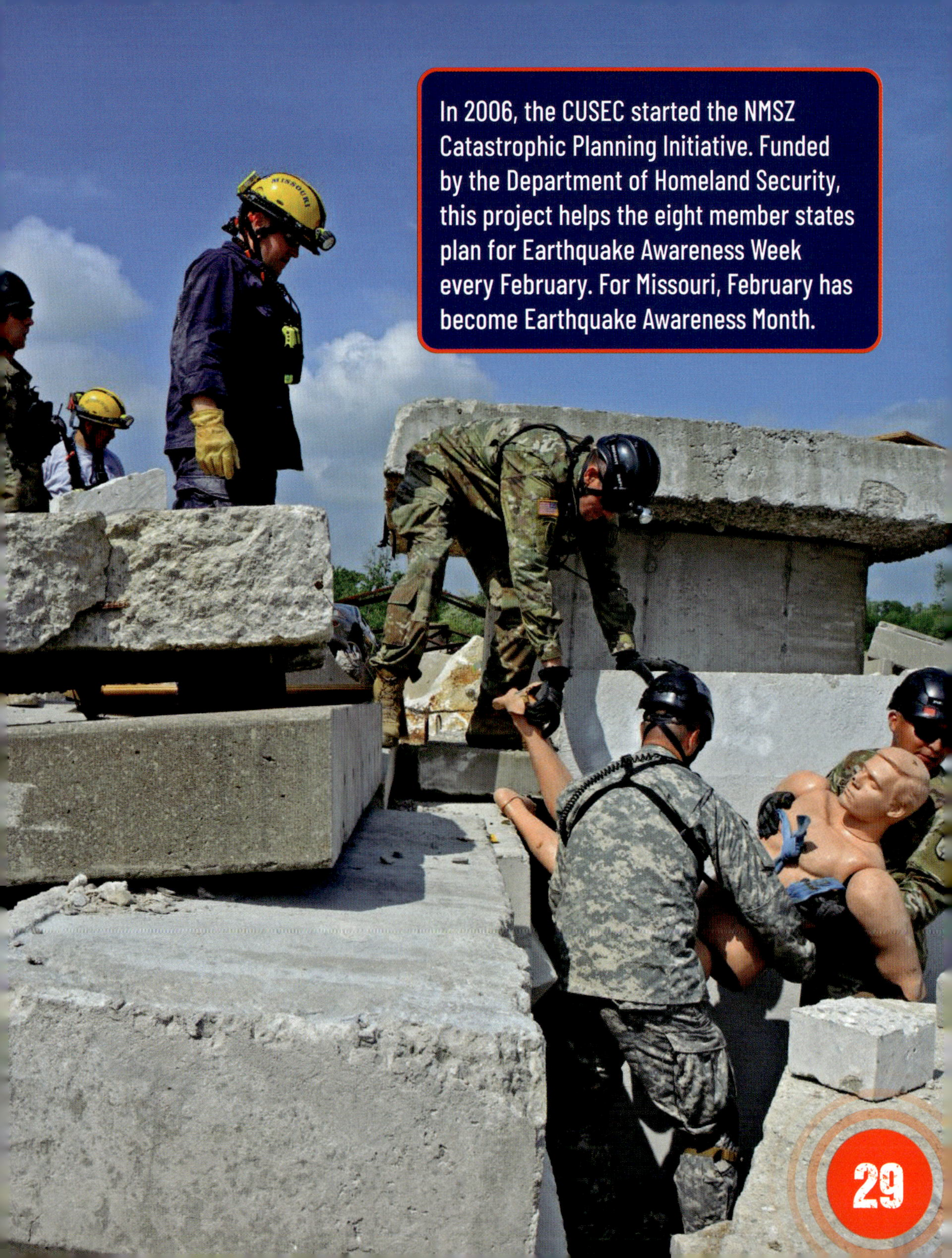

In 2006, the CUSEC started the NMSZ Catastrophic Planning Initiative. Funded by the Department of Homeland Security, this project helps the eight member states plan for Earthquake Awareness Week every February. For Missouri, February has become Earthquake Awareness Month.

GLOSSARY

accumulation: Describing something that has built up over time.

aftershock: A smaller earthquake that follows the main shock of a larger earthquake.

alluvial: Having to do with land created by sediment transported by river water.

diverse: Having many different types or forms.

dormant: Not active for the time being.

embed: To enclose or encase something in a surrounding mass.

epicenter: The part of the Earth's surface directly above the area where an earthquake occurs.

geyser: A jet of water that shoots out of the ground.

levee: A bank built along a river to stop flooding.

magnitude: Size or extent.

phenomenon: A fact or event that is observed.

sediment: Matter such as rocks, sand, and stones that's moved and deposited by water, wind, or glaciers.

significant: Noticeably large or important.

supercontinent: A former large continent that is assumed to have existed and from which other continents broke off and drifted away.

tremor: A trembling or shaking of the ground.

FOR MORE INFORMATION

BOOKS

Doyle, Abby Badach. *Plate Tectonics Reshape Earth!* New York, NY: Gareth Stevens Publishing, 2021.

Edwards, Jamee-Marie. *Earthquakes and the Environment*. North Mankato, MN: Capstone Press, 2025.

Izzi, Howell. *Earthquakes*. Buffalo, NY: Enslow Publishing, 2025.

WEBSITES

Earthquakes for Kids
earthquake.usgs.gov/learn/kids
Read much more about the science of earthquakes, Earth's plates, and see amazing photographs of earthquake damage as well as scientists hard at work.

New Madrid Historical Museum
www.newmadridmuseum.com
Learn more about the history of New Madrid. Information includes the earthquakes but also Native Americans, the Civil War, and more.

INDEX